cano

Iceland's
Inferno
and Earth's Most
Active Volcanoes

With strong winds
blowing heat and gases
away from him, photog-
rapher Carsten Peter
worked in respirator
and helmet a mere 20
feet from this turbulent
eruption of Mount Etna,
in Sicily, in 2001.

Volcano

Iceland's Inferno
and Earth's Most Active Volcanoes

Edited by Ellen J. Prager

Foreword by Marcia K. McNutt
Director, U.S. Geological Survey

NATIONAL GEOGRAPHIC

WASHINGTON, D.C.

This night view captures
the 1989 eruption of
Nuamuragira, in Zaire's
Virunga Mountains,
part of Africa's Great
Rift Valley.

COVER
Lava fountains spew
dramatically during the
March 2010 eruption
in Iceland at Fimmvor-
duhals, a ridge near
Eyjafjallajokull glacier.

Contents

Foreword

by Marcia K. McNutt, Ph.D., Director, U.S. Geological Survey

Few natural phenomena are as spectacular as a volcanic eruption: incredible energy bursting from the Earth with such intense heat that rock is reduced to flowing liquid. Millions of people have marveled at the ruins of Pompeii, its remarkably advanced civilization frozen in time over 19 centuries ago by Mount Vesuvius, standing calmly in the distance with modern Naples spread out at its feet. Earth, however, is just as restless today as it was in Roman times, and our civilization is in some ways much more at risk.

As recently as April 2010, a small eruption in Iceland that might otherwise have escaped notice drew the attention of the world by stopping air traffic over Europe for five days because of the vulnerability of our technology to volcanic ash. Combined with the exponential expansion of the human population since A.D. 79, our exposure to natural disasters in general and volcanic eruptions in particular is extraordinary. A major eruption of Vesuvius now would be as costly to life as a war if people were not evacuated in time.

The good news is that most eruptions can now be forecast to save lives and property, provided that volcanoes are properly monitored and there are emergency plans in place. Reading the seismic signs of imminent eruption allowed the Icelandic government to evacuate people living in the danger zone. Lives were saved from the much larger eruption of Kasatochi volcano in the Aleutian Islands in 2008, when the U.S. Geological Survey's Alaska Volcano Observatory recognized impending disaster from an intensifying storm of earthquakes. Airplanes en route between Asia and North America were able to safely avoid the ash cloud that quickly reached flight levels. This was a lesson first learned in 1989, when a Boeing 747 with 245 people aboard lost power in all four engines after flying into an eruption cloud from Mount Redoubt. The plane dropped nearly 14,000 feet before the pilot could restart two engines and land safely in Anchorage.

Technological developments now permit detection of the slight swelling of a volcano prior to eruption, both from ground stations and from orbiting satellites. Seismometers can pick up the earliest drumbeats that signal magma on the move. Satellites can distinguish among rain clouds, ash clouds, and volcanic gas clouds, as well as detect surface "hot spots." These observations are made useful to the public by new technologies for displaying and transmitting data, and a new commitment by everyone—scientists, governments, and the public alike—to plan ahead.

In the United States, efforts to reduce vulnerability to volcanic hazards were jump-started by the dramatic explosion at Mount St. Helens 30 years ago. The volcano was like a flame drawing moths—scientists and citizens—to potential danger and in some cases death. Two years before, U.S. Geological Survey experts had identified it as the most active volcano in the Cascade Range and suggested that an eruption might occur before the end of the century. Yet few American geologists believed they would actually see a mainland eruption in their lifetimes. That changed at 8:32 a.m., May 18, 1980, with the deadly blast that took 57 lives.

This smoking cone of ash began growing from the remains of Indonesia's Krakatau in 1927. Scientists expect Anak Krakatau, "child of Krakatau," to someday erupt violently—as its parent did in 1883.

Since that time, the USGS has established four new volcano observatories and partnered with the U.S. Agency for International Development to create an international eruption response team that assists developing countries with volcanic crises. Domestically, USGS has tracked eruptions in Alaska and Hawaii, volcanic unrest in Long Valley Caldera, California, and Yellowstone Caldera in Wyoming, and the sudden Mount St. Helens reawakening from 2004 to 2008. USGS volcanologists and their colleagues have learned where magma is stored beneath volcanoes, how it ascends and explodes, and where ash clouds and pyroclastic, lava, and mud flows are likely to go.

There remains a challenge for all governments and communities that have volcanoes in their neighborhood: maintaining the vigilance and investment required to protect lives and property, even though an individual volcano, like Vesuvius or St. Helens, may slumber for lifetimes. With 169 potentially active volcanoes in the United States and its territories, that vigilance is clearly a necessity. Volcanic eruptions have been a part of this planet for billions of years. We can prevent these natural hazards from becoming human catastrophes by investing in science and preparation before the sleeping giants awake.

Belching volcanoes such as Mount Semeru, background, and Mount Bromo, far left, are portals to a subterranean world that shapes not only Indonesia's landscape, but also its beliefs and culture.

About 1,900 active volcanoes can be found on Earth.

Vanuatu's Marum
Volcano spews fire-
works as a volca-
nologist inches along
a crust of rock. "You
have to be addicted to
volcanoes to take that
risk," says photogra-
pher Carsten Peter.

Full helmet and thermal suit—and years of learned caution—allow French technician Charles Rivière to stay relatively safe as he gathers lava samples close to the action on Mount Etna in 2001.

"It's a fantastic force of nature, with terrific noise and a very real danger of getting hit."

Charles Riviere, research technician

They form some of the world's most majestic features even when standing cold and silent, but when subterranean events converge to set them off, they become nothing short of astonishing. "It's almost like a religious experience," says volcanologist Haraldur Sigurdsson. "It's both thrilling and terrifying to realize that we don't know when the next big eruption will occur." Add to that the potential to destroy entire communities and ecosystems in minutes, to disrupt world travel and events, and to remind us, uneasily, of the violent forces lying just beneath our feet. On Earth, volcanoes are not the exception, they are the norm.

Well into the Middle Ages, many people thought of volcanoes, with their fiery summits and unearthly roarings, as the hellish world of suffering sinners. As science evolved, so did our understanding of the forces underlying volcanoes, which cause them to explode with terrifying power, spew fountains of lava, release deadly flows of mud and

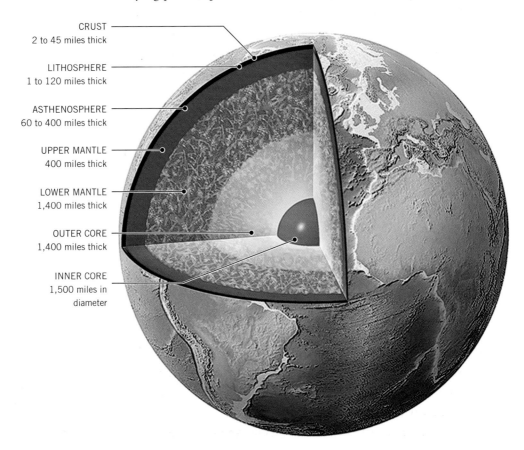

CRUST
2 to 45 miles thick

LITHOSPHERE
1 to 120 miles thick

ASTHENOSPHERE
60 to 400 miles thick

UPPER MANTLE
400 miles thick

LOWER MANTLE
1,400 miles thick

OUTER CORE
1,400 miles thick

INNER CORE
1,500 miles in diameter

On Guatemala's Pacaya volcano a lava skylight—an opening in the roof of a lava tube—is a portal to the volcanic fury that lies within.

debris, and send towering plumes of superheated gas and ash into the stratosphere. Or sometimes, all of the above.

The Earth's surface and everything on it, including volcanoes themselves and the land shaped by their violent outpourings, are just a thin veneer enveloping the planet. At its center is a metallic core—a ball of iron and nickel alloy hotter than the surface of the sun that is kept solid by the tremendous pressure around it. Surrounding the solid core is a layer of liquid metal whose flow in the rotating Earth generates electricity, creating the planet's magnetic field. Atop this layer is a thick mantle of rock that churns ever so slowly, driven by high temperature and pressure. Encasing the whole is the thin rigid crust of the Earth, which includes all of the seafloor and continents. Within the mantle, plumes of hot molten rock rise toward the surface, cool, and sink—an internal plumbing system that has cracked the thin crust into a jigsaw puzzle of rocky slabs. These tectonic plates may measure thousands of miles across but are comparatively thin—45 miles or less—about the same in relation to the Earth as an eggshell to an egg. The plates move independently a few inches a year. At their edges, they grind against each other, collide, and spread apart, triggering earthquakes and causing eruptions.

Volcanoes are not randomly distributed over the Earth's surface but are concentrated at the edges of tectonic plates or at hot spots on the plates. Where two plates collide, huge slabs of crust can be thrust downward, where intense heat causes melting. Hot, molten magma becomes more buoyant than the surrounding rock and rises

On the Big Island of Hawaii, water turns to steam as molten lava from Kilauea hits the ocean. All seven Hawaiian islands are formed by volcanism, but only the Big Island's Mauna Loa and Kilauea are currently active. Around the world, eruptions tear down and rebuild the landscape.

up through the overlying slab and crust. Where the two plates have collided, a nearby chain of volcanoes is built by the eruption of the melted, recycled magma from the underlying slab. The volcanoes of the Andes and the Pacific Northwest's Cascade Range were born of plate collisions and remain potentially active and explosive.

Where plates move apart they create another class of volcanoes, best exemplified by the oceanic ridge system, a 50,000-mile-long undersea mountain chain encircling the Earth like a giant zipper. Here, as the plates spread apart, magma erupts and creates new seafloor in the form of rocky ridges, steep cliffs, and wide underwater valleys. The Mid-Atlantic Ridge actually breaches the surface in volcanically active Iceland and in the Azores. Volcanoes also form at hot spots, where plumes of magma rise through the mantle to the surface. Chains of volcanic islands can be created, such as those that make up Hawaii and the Galapágos Islands. From modest heaps of cinder to towering cones

Volcanism on Earth occurs in three main settings: at subduction zone, near right, at spreading centers, middle, and over hot plumes of material that rise from deep in the interior, far right.

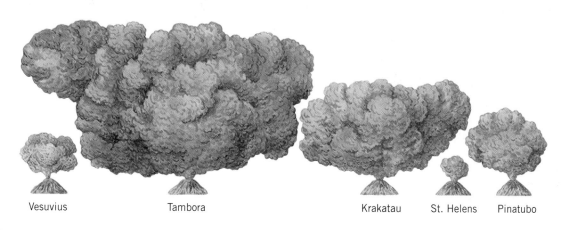

Vesuvius   Tambora   Krakatau   St. Helens   Pinatubo

whose snowy summits pierce the clouds, volcanoes vary tremendously in shape, size, and behavior. A volcano's form is generally determined by the size and type of its eruptions, which are largely dependent on the chemical composition of the underlying magma. Like individuals with unique quirks of personality, volcanoes exhibit distinct characteristics. Steaming fumaroles and bubbling mud pits may be telltale signs of the activity below, or a volcano may be a sleeping giant, remaining dormant for long periods between eruptions.

Around the world, there are hundreds of potentially active volcanoes that put millions of people at risk. While we can't know with certainty when or where the next volcanic catastrophe will occur, we do know it's only a matter of time.

In 1815 Indonesia's Mount Tambora ejected so much ash that it temporarily changed Earth's climate. It spewed out four times more debris than Krakatau in 1883 and 80 times more than Mount St. Helens in 1980.

Volcanoes of the World

EURASIAN

PLATE

Redoubt
Volcano

Mt.
Katma

JUAN DE FUCA
PLATE

Mutnovsky

HIMALAYA

Plateau
of Tibet

Mt. Unzen

Mt. Fuji: World's largest composite
(explosive) volcano; 35 million
people live in its shadow.

Hawaiian Islands

ARABIAN
PLATE

INDIAN

PLATE

Dallol

Mt. Pinatubo: Eruption in 1991 is
the largest eruption in living memory;
its ash reduced temperatures in the
northern hemisphere by nearly 1°F.

PHILIPPINE
PLATE

Mauna Loa: World's largest
active volcano; rises 30,000 feet
from ocean floor to summit.

Kilaue

PACIFIC

PLATE

Mt. Toba: Eruption 74,000 years
ago was largest in human history;
likely reduced the human population
to fewer than 10,000 individuals.

AFRICAN

PLATE

Krakatau: 1883 eruption was so
loud it was heard 3,000 miles away;
subsequent tsunamis killed 36,000.

Semeru

Mt. Tambora: 1815 eruption
was the largest and deadliest in
recorded history, killing 92,000.

Marum

AUSTRALIAN

PLATE

Location Uncertain

Plate Boundaries

- Divergent boundary
- Convergent boundary
- Transform zone
- ◯ Hot spot

ANTARCTIC

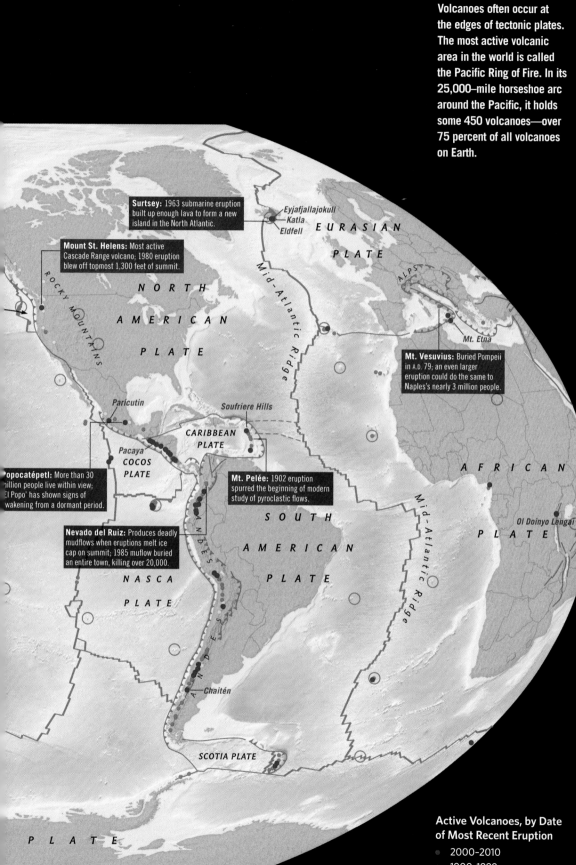

Volcanoes often occur at the edges of tectonic plates. The most active volcanic area in the world is called the Pacific Ring of Fire. In its 25,000-mile horseshoe arc around the Pacific, it holds some 450 volcanoes—over 75 percent of all volcanoes on Earth.

Surtsey: 1963 submarine eruption built up enough lava to form a new island in the North Atlantic.

Eyjafjallajokull
Katla
Eldfell

EURASIAN
PLATE

Mount St. Helens: Most active Cascade Range volcano; 1980 eruption blew off topmost 1,300 feet of summit.

ROCKY MOUNTAINS

NORTH

AMERICAN

PLATE

Mid-Atlantic Ridge

ALPS

Mt. Etna

Mt. Vesuvius: Buried Pompeii in A.D. 79; an even larger eruption could do the same to Naples's nearly 3 million people.

Paricutin

Soufriere Hills

Pacaya
Popocatépetl: More than 30 million people live within view; 'El Popo' has shown signs of awakening from a dormant period.

COCOS
PLATE

CARIBBEAN
PLATE

Mt. Pelée: 1902 eruption spurred the beginning of modern study of pyroclastic flows.

AFRICAN

Ol Doinyo Lengai

Great Rift Valley

PLATE

SOUTH

Nevado del Ruiz: Produces deadly mudflows when eruptions melt ice cap on summit; 1985 muflow buried an entire town, killing over 20,000.

ANDES

AMERICAN

PLATE

Mid-Atlantic Ridge

NASCA

PLATE

Chaitén

SCOTIA PLATE

PLATE

Active Volcanoes, by Date of Most Recent Eruption

- 2000-2010
- 1900-1999
- 1500-1899

tallest active volcano
mt. etna
10,925 ft

deadliest eruption
vesuvius
A.D. 79

tectonic location
eurasian and
african plates

The volcanic lightning
storm has a makeup
similar to a regular
thunderstorm.

ICELAND Eyjafjallajokull

The aurora borealis
dances above the ash
plume of Iceland's
Eyjafjallajokull volcano
during the evening
hours of April 22, 2010.

One-third of all the lava that has erupted from the Earth in the past 500 years has flowed out of Iceland.

Scientists race from Eyjafjallajokull's eruption site in eastern Iceland on April 15, 2010, after collecting samples of ash for analysis.

stats
Iceland: Eyjafjallajokull

Type of volcano:
Stratovolcano

Last significant eruption date:
2010

Elevation above sea level:
5,466 feet

Status:
Historical

Volcanic alert:
After nearly 200 years of dormancy, Eyjafjallajokull, one of Iceland's largest volcanoes, exploded, sending up a 36,000-foot-high ash cloud that caused most of Europe's airports to close for days.

An aircraft with it engine protected is grounded in Belfast, Northern Ireland, opposite above. On April 19, five days after Eyjafjallajokull's first eruption, passengers from India stranded in Munich, Germany, wait for air traffic to resume.

After two centuries of rest, one of Iceland's largest volcanoes came rumbling back to life in 2010, causing the worst peacetime disruption to air travel in history. Long-known as the land of fire and ice, Iceland has a history of spectacular eruptions, but no one seems to have been prepared for the problems brought on by the 2010 blast.

The Eyjafjallajokull volcano started showing signs of activity around March 20, when a glowing red cloud rose over the underlying glacier. Lava began to fountain from volcanic vents and flow in thick rivers down the mountain's northeast side. Then the activity quieted and calm returned—until an explosive blast on April 14 sent gas and ash thousands of feet into the air. As glaciers melted and the threat of flash floods loomed, nearby residents were evacuated to safety, but the danger was expanding.

High in the atmosphere, a plume of volcanic ash was spreading quickly. Soon air traffic in northern and central Europe was grounded and airports closed. Although no one knew exactly what concentrations of ash were present or how much it would take to cause a problem, nobody wanted to take a chance. Volcanic ash clouds have previously damaged jet engines and caused them to shut down. Millions of passengers were stranded as flights were canceled, costing the airline industry over three billion dollars.

The last eruption of Eyjafjallajokull occurred in 1821, covering much of Iceland in a dark blanket of ash, and lasted for almost a year. It was followed in 1823 by the eruption of a neighboring and larger volcano, Katla. As the 2010 eruption seemed to subside and global air travel resumed, scientists continued to monitor Eyjafjallajokull and watch for a rekindling of activity or the possibility of a related, potentially larger eruption on Katla.

Straddling the Mid-Atlantic Ridge and sitting atop a hot spot, Iceland is one of the most volcanically active regions in the world, an eerie assemblage of volcanic landforms, geysers, ashy plains, smoky bays, glowing fissures, and fire-breathing mountains. For Icelanders, volcanic eruptions are a way of life, a part of their history. For the rest of the world, the recent eruption of Eyjafjallajokull has brought new attention to the power and danger that lurks within the world's sleeping volcanoes.

With Eyjafjallajokull's
eruption still going
strong on April 17,
farmer Thorarinn
Olafsson tries to lure
his horse to the stable
as a cloud of black
ash looms overhead.

"... the whole island seemed one globe of fire, and the sea on every side boiled up ... like a caldron set on a fire ..."

St. Brendan the Voyager, sixth century A.D.

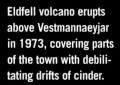

ICELAND Eldfell

Eldfell volcano erupts above Vestmannaeyjar in 1973, covering parts of the town with debilitating drifts of cinder.

n 1973 a massive eruption on the small island of Heimaey forced the evacuation of more than 5,000 people from the town of Vestmannaeyjar, one of Iceland's most important fishing ports, causing a serious threat to the economy. Spewing smoke, fire, and lava bombs, Eldfell ("fire mountain") erupted with a thunderous blast on the morning of January 23. Boiling rivers of lava and hot ash engulfed a third of the town and threatened to seal the harbor. Jetting lava reached twice the height of the Empire State Building as night turned from black to red. An eerie glow was visible from 50 miles out at sea. By the end of two weeks, a 700-foot-high black volcanic cone loomed over the town. Desperate islanders turned fire hoses on the mountain as it crawled closer. The effort paid off. They managed to slow the advance of the lava and divert it away from the town and into the sea. When it was all over, miraculously only one death was reported.

Vesuvius

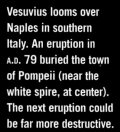

Vesuvius looms over
Naples in southern
Italy. An eruption in
A.D. 79 buried the town
of Pompeii (near the
white spire, at center).
The next eruption could
be far more destructive.

stats
Italy:
Vesuvius

Type of volcano:
Somma volcano

Last significant
eruption date:
1944

Elevation above
sea level:
4,203 feet

Status:
Historical

Volcanic alert:
After lying dormant
for centuries, Vesuvius
awoke without warn-
ing in A.D. 79, spewing
volcanic ash and dust
that buried the Roman
towns of Pompeii
and Herculaneum.
Today more than two
million people live
near the volcano.

Body casts of an adult
and children lie where
they were struck down
by an asphyxiating
mixture of gas and ash
from Mount Vesuvius in
A.D.79.

One of the world's most dangerous volcanoes looms large over the lives of more than two million people in southern Italy. An eruption in A.D. 79 famously buried the towns of Pompeii and Herculaneum, taking thousands of lives, and new research suggests that the next blast of Vesuvius could be even bigger. In 2006, scientists published the results of a decade-long study that received worldwide attention and was a dire warning for the region, especially the city of Naples. The 79 eruption, it seems, had been preceded by a far more destructive event that could someday be repeated.

Dubbed Avellino, the eruption occurred some 3,780 years ago. The blast hurled nearly 100,000 tons a second of superheated rock, cinders, and ash into the stratosphere, reaching an altitude of more than 20 miles. Within 12 hours, evidence suggests, the towering column of ash collapsed, producing an apocalyptic sequence of events that makes this type of volcanic event one of the most lethal natural hazards on Earth.

The collapsing ash cloud creates a pyroclastic surge—a searing, turbulent avalanche of debris that shoots out sideways from the slopes of a volcano. The scorching cloud can travel at incredible speeds, over 200 miles an hour, and reach temperatures nearing 1000°F. After Vesuvius's Avellino eruption, ash and soot deposits heaped as deep as 65 feet three miles from the volcano's crater and nearly a foot thick 15 miles away. (Eight inches of ash is enough to cause modern roofs to collapse.) The pyroclastic cloud that devastated the region was followed by massive volcanic mudflows that buried or swept away whatever was left. The Avellino event lasted less than 24 hours and left the once idyllic landscape uninhabitable for some 300 years.

Nothing approaching that cataclysm or the A.D. 79 eruption has occurred since, but the deadly mountain has left its mark in more modern times. In 1631 a glowing avalanche killed 3,500 people, and in 1944 gas and ash shot skyward and later overran several villages, killing dozens and damaging 88 Allied aircraft based nearby. In recent decades Vesuvius has stayed relatively quiet, and streams of tourists clamber up the steep path to its crater. Few of them pause to think about the vast reservoir of magma that lies six miles below. Vesuvius is inarguably a ticking time bomb; we know that catastrophic eruptions have occurred about every 2,000 years. Unfortunately, the next one may be coming due.

Vesuvius is a ticking time bomb; the current evacuation plan covers the 600,000 people who live in the Red Zone.

IMPACT OF A MAJOR VESUVIUS ERUPTION
Risk zones based on 1780 B.C. eruption

People indoors would likely survive; most trees toppled.

Little chance of human survival; most buildings damaged

100 percent mortality; most buildings destroyed

Total devastation

Afragola

NAPLES MUNICIPALITY BOUNDARY

Naples

Maschio Angioino

Portici

Herculaneum   Ercolano

ITALY

Apennines

Rome ★

Adriatic Sea

Vesuvius

Naples

Capri →   ← Sorrentine Peninsula

40°N

Tyrrhenian Sea

0 mi   150
0 km   150

Sicily

Shadowing the ruins of Pompeii, Vesuvius buried this ancient town under 15 to 25 feet of ash in A.D. 79.

OFFICIAL EVACUATION AREA
Municipal boundaries define the high risk area the government calls the Red Zone.

Castel Cicala
Nola

San Paolo
Bel Sito

C a m p a n i a

The ancient Monte Somma crater will funnel pyroclastic surge toward Naples.

Vesuvius
4,203 ft
1,281 m

Pompeii

Gulf of Naples

N

SCALE VARIES IN THIS PERSPECTIVE.
DISTANCE FROM VESUVIUS TO MASCHIO ANGIOINO IS 9 MILES (15 KILOMETERS).

SOURCES: GIUSEPPE MASTROLORENZO AND LUCIA PAPPALARDO, OSSERVATORIO VESUVIANO, NAPLES; PIER PAOLO PETRONE, MUSEUM OF ANTHROPOLOGY, UNIVERSITY OF NAPLES; MICHAEL SHERIDAN, UNIVERSITY AT BUFFALO, NEW YORK; LANDSAT IMAGE BY PROSPECTRA

JEROME N. COOKSON AND LISA R. RITTER, NGM MAPS

During World War II, Vesuvius shot a roiling cloud of ash skyward as U.S. bombers flew on a mission. This eruption destroyed villages, killed dozens of people, and damaged 88 Allied aircraft on the ground.

Etna

For 24 days in the summer of 2001, Mount Etna put on a spectacular show. Monitored by scientists, rivers of lava and fountains of fire awed residents near the famed Sicilian peak.

Mount Etna's name comes from the Greek *aitho,* meaning "I burn."

The explosive birth of
a cone 6,900 feet up
Etna's southeast flank
hurls ash and hot rock.
Sound waves from the
blast rattled windows
20 miles away.

"You have to stand still, watch where the lava bombs are falling, and get out of the way."

Carsten Peter, National Geographic photographer

stats
Italy: Etna

Type of volcano:
Stratovolcano

Last significant eruption date:
2009

Elevation above sea level:
10,925 feet

Status:
Historical

Volcanic alert:
Mount Etna is considered the most active volcano in the world and has the longest recorded history of eruptions, some 200 going back to 1500 B.C.

Scientists watch as a massive red cloud of roiling volcanic ash and gas—reaching nearly a mile high—blasts from a fissure on the flank of Mount Etna.

In 2001 Mount Etna put on the most dazzling show of a decade. For 24 days during the summer, rivers of lava and fountains of fire awed even those who had long lived in the shadow of Sicily's famed volcano. During the eruption sound waves rattled windows nearly 20 miles away, and exploding ash and scorching rocks built a cone 6,900 feet high on the mountain's southeast flank.

Mount Etna dominates northeastern Sicily. Towering nearly 11,000 feet, it is Europe's highest active volcano and has long lured volcanism voyeurs. Plato sailed from Greece in 387 B.C. just to have a look at it, and legend tells of Odysseus's battle with a Cyclops on the volcano's flanks. Romans considered Etna to be Vulcan's forge, and ninth-century Arabs created sweet, flavored ices from its snow. Today, smoke emanating from the volcano's peak continues to be evidence of the heat and active magma below, but because eruptions occur at upper elevations and its lava creeps slowly, Etna rarely takes human lives.

In late June of 2001, Etna's activity seemed routine; slender tornadoes of gas spewed from the crater's summit. But then unusual blasts rocked the crater, and a fissure opened along its side, gushing lava. Five new gashes opened up farther down its flank and began spilling lava as well. With breathtaking speed, two cones rose on the volcano's side, one spouting lava fountains as tall as 1,300 feet.

People living within Etna's reach had to deal with the eruption's impact. They brought in bulldozers to build an embankment and divert potentially dangerous flows away from a popular ski-lift base and scientific monitoring station. The lava eventually stopped less than three miles from Nicolosi, the most threatened of the region's small towns. Ashfall closed the Catania airport for several days and some roads and property were destroyed, but overall Etna kept its local reputation as a "friendly giant."

The same volcanic eruptions that can destroy homes and property and cause economic and travel woes can also bring life and fertility. The region's ash-enriched soil supports productive vineyards and lush orchards. People in towns along Etna's lower slopes accept the land's dangers as a trade-off for its bounty. Since 2001 the volcano has continued to erupt annually, bringing lava flows and explosive blasts but sparing local communities.

About 1 in 12 of the world's population now live around Earth's active or potentially active volcanoes.

The lights of Sicily's Catania and the Ionian Sea coast spread below the new Piano del Lago cone. The ancient name Etna

tallest active volcano
mt. meru
14,977 ft

deadliest eruption
mt. nyiragongo
1977

tectonic location
african plate

The continuously erupting Erta Ale is not only the most active volcano in Ethiopia, but it lies in an area below sea level, making it one of the lowest volcanoes on Earth.

Glowing lava erupts
from a hornito in
Tanzania's Ol Doinyo
Lengai crater. The
otherworldly landscape
is the result of slick
sodium carbonate lava,
which has roughly the
same consistency as
olive oil.

Ol Doinyo Lengai

Unusually cool and
highly fluid lava flows,
which have a chemical
composition akin to
laundry soap, produced
this foot-long wing,
which froze in midair
and then shattered
within 48 hours.

"It's absolutely incredible. Like being on the moon."

Ciela Nyamweru, geographer, St. Lawrence University

Dallol
AFRICA
Great Rift Valley
Lake Victoria Equator
Lake Natron
Ol Doinyo Lengai
TANZANIA

stats
Tanzania: Ol Doinyo Lengai

Type of volcano:
Stratovolcano

Last significant eruption date:
2010

Elevation above sea level:
9,718 feet

Status:
Historical

Volcanic alert:
After decades of low activity, Ol Doinyo Lengai erupted explosively from September 2007 to April 2008, showering ash on nearby villages. Thousands of nomadic herdsmen had to temporarily evacuate because the grasslands where their cattle grazed were covered in ash.

Hardening within seconds of extrusion, volcanic froth rich in carbon dioxide spews from an active vent. What begins as liquid lava hits the ground with the tinkling of breaking glass.

To the local Masai, Ol Doinyo Lengai or "Mountain of God" is sacred, a place for pilgrimage and worship. To geologists it is one of the strangest volcanoes on the planet.

Lengai is the world's only active carbonatite volcano: Its lava erupts only half as hot as other volcanoes and is made mostly of sodium carbonate, chemically akin to laundry soap. When he visited the volcano in the late 1980s, famed French volcanologist Maurice Krafft memorably counseled his companions: "Remember, this lava is cooler than most. If you punch through, you'll have time to scream before you die."

Ol Doinyo Lengai sits in northern Tanzania, part of the Great Rift Valley. The current summit rises 7,650 feet above the parched valley floor, overshadowing an inactive southern crater that dates from when most of the volcano was formed about 15,000 years ago. The volcano erupted violently in 1966 and continues to rumble and spew lava, creating what has been described as a Dr. Seussian world of geology.

The cool acidic lava is highly fluid, roughly the same runniness as olive oil. When it erupts, it is richly black but quickly transforms within a few days into muddy brown, gray, and finally a frosty white. Collapsed spatter cones (formed when gobs of ejected lava pile up around a vent) exude rivers of new lava, looking for all the world like streams of melted chocolate. It hardens fast when exposed to the atmosphere and becomes brittle, decaying quickly—even raindrops leave marks. This rapid hardening is not limited to liquid lava. Silvery volcanic froth rich in carbon dioxide that sprays from active vents turns into strange air-borne sculptures as it hardens in mid-flight. Solidified lava fountains become bizarre towers that resemble enormous versions of a child's drip sand castle. And lava flows strangely similar to those found on the other side of the planet in Hawaii, some sharp, some ropy, pave the landscape.

For good or ill, the unusual chemical nature of the volcano has helped to shape the surrounding lands. Ash from Ol Doinyo Lengai has left the lakes to the north strongly alkaline. It also has blanketed the Serengeti, which stretches out beyond the volcano's western flank. Rain turns the plain's ash into a hardpan impenetrable to tree roots, leaving a landscape where only shallow-rooted grasses are able to take hold.

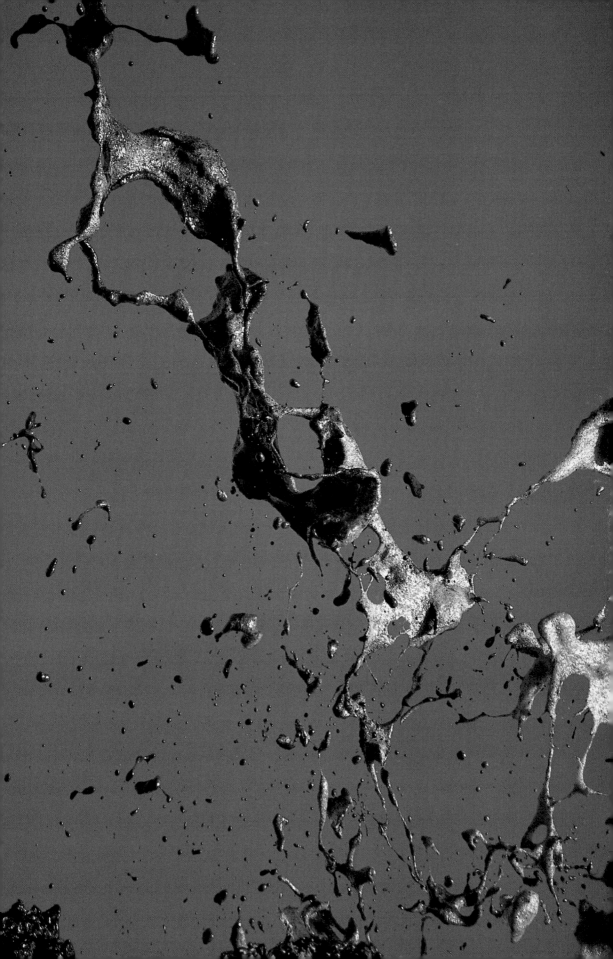

On a moonlit night atop Ol Doinyo Lengai, photographer Carsten Peter held the shutter of his camera open as lava surged from an erupting hornito, or mound. He captured the lava's glow, hidden to the naked eye.

Great Rift Valley

Four-million-year-old hominid fossils are preserved by the Great Rift Valley's volcanic ash.

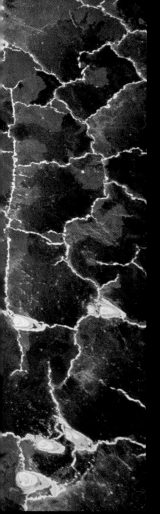

Visitors to Kenya know the Great Rift as the breathtaking escarpments they pass on safari. Few people realize it is actually an immense series of fissures that are slicing the African continent apart. It is an opening wound that runs for nearly 4,000 miles from the Red Sea south to Mozambique. Enormous cracks—in places a mile deep and 50 miles across—have opened up along its length. In Central Africa the rift has two branches: The Eastern Rift Valley bisects Kenya and skirts both Kilimanjaro and the Serengeti Plain in Tanzania; the Western Rift Valley cleaves the heart of the continent, cupping a great chain of lakes. The tearing earth lifts the Ruwenzori Massif and stokes the volcanic fires of the Virunga Mountains, home to the endangered mountain gorilla. Lake Victoria sits atop a plateau between the two branches. The Great Rift began to open some 30 million years ago, but the process has by no means ended. Experts believe we could be witnessing the first stages in the development of a new ocean basin.

An airplane casts a shadow over Tanzania's Lake Natron, above. The lake's color comes from a red pigment in cyanobacteria. In Dallol Volcano, upper right, an Afar Tribesman tests the water of a hot spring. Discs of travertine ring a 12-foot-wide hot spring in the Danikil Depression.

Great Rift Valley

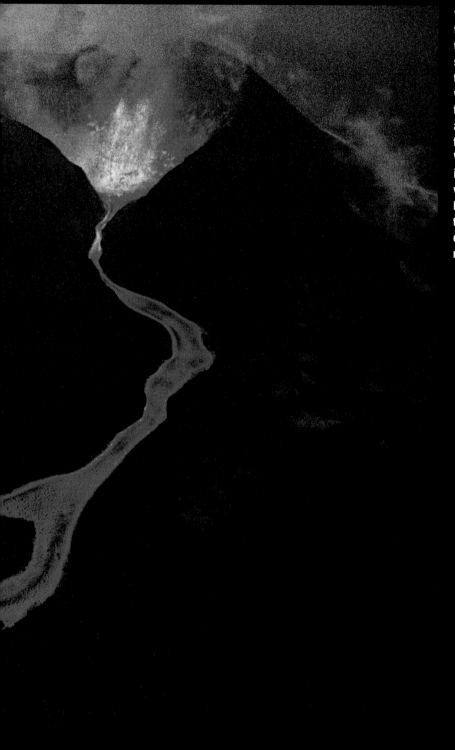

"It sounded like a freight train rumbling through a thunderstorm," recalls photographer Chris Johns, who witnessed this fissure spewing lava and molten rivers 98 feet across in Zaire's Virunga Mountains in May 1989. Such violent eruptions are a by-product of the rifting that has been ripping eastern Africa apart for millions of years.

tallest active volcano
mauna loa
13,681 ft

deadliest eruption
lamington
1951

tectonic location
australian and
pacific plates

In all its neon-colored-glory, New Zealand's Champagne Pool hot thermal spring lies in the heart of a fermenting volcanic landscape. Carbon dioxide gases in the boiling water make the pool bubble like its namesake.

VANUATU **Marum**

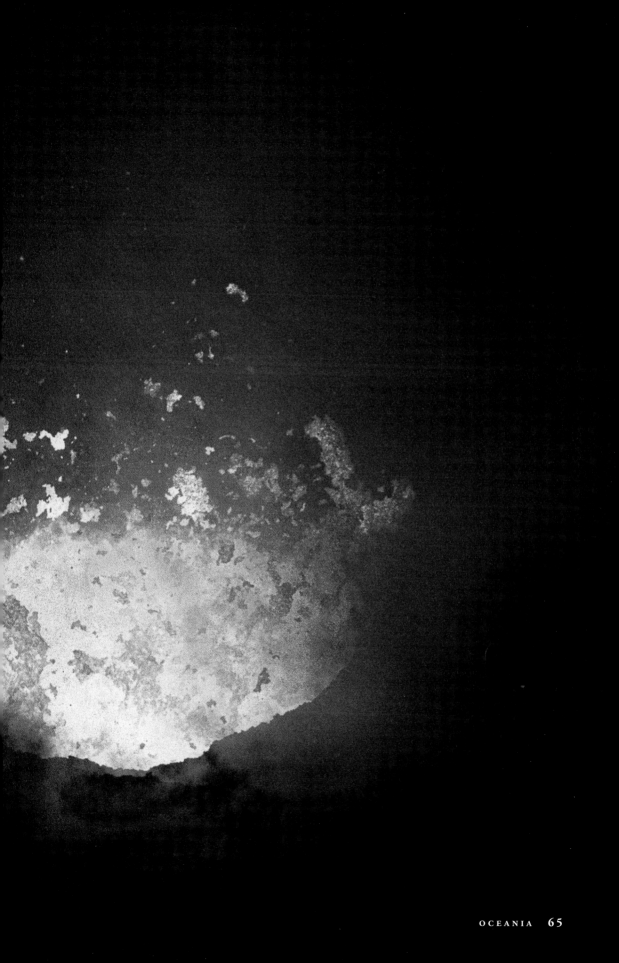

Veteran volcano
expert Franck Tessier
descends into Marum,
coming within 150
feet of lava as hot as
2200˚F. He and his
wife, Irène Margaritis,
have gone deeper
into this crater than
anyone before.

Some volcanic eruptions have a force greater than the most powerful nuclear bomb.

Pacific Ocean

SOLOMON ISLANDS

New Guinea

Mount Marum

VANUATU

Coral Sea

AUSTRALIA

Tropic of Capricorn

stats
Vanuatu: Marum

Type of volcano:
Shield volcano

Last significant eruption date:
2010

Elevation above sea level:
4,167 feet

Status:
Historical

Volcanic alert:
One of the most active volcanoes in the region, eruptions have been recorded here nearly every year since historical documentation started.

The islands of Vanuatu range from lush to deadly. Geared up for the dangers of an active crater, on the island of Ambrym, German engineer Chris Heinlein displays one of the many insects lured by the glow of lava—and doomed to die in the fumes or heat. Back at base camp, Heinlein goes safely barefoot in the ash.

One of the most active volcanoes in the South Pacific, Marum sits on the island of Ambrym among the 80 islands that make up the nation of Vanuatu. In Marum's crater, gases burst to the surface within a roiling lava lake, launching molten bombs and creating noxious clouds. Surrounding the volcano is a vast ash plain, seven miles across and hundreds of feet thick. It is a desolate volcanic entryway to the mountain's hellish peak, some 4,000 feet high.

Partway up the volcano's slope is Niri Taten, a smaller crater that tunnels straight down into the steaming basaltic rock, like a giant burrowing worm with steaming breath. Near the crater's edge, 50-mile-an-hour winds whip up bits of stone and grit, as thick clouds of gas roar upward. The rocky surfaces within Niri Taten are painted in spectacular colors. Some of its sheer rock faces are swathed in bright yellow, covered with sulfur. Iron washes other sections of rock in flaming orange. Pastel green deposits of manganese glaze rock nearest the vent, and other patches of stone are bleached white by chlorine and fluorine gases.

The summit crater is a wide-open volcanic pit, stepped by ledges and marbled by layers of black ash and pale bleached rock. The crater is 1,200 feet deep, born in a great blast that tore through layers of hardened volcanic rock. Small wall vents called fumaroles—created where heated groundwater and escaping gases breach the surface—let off steady plumes of steam. At the bottom of the crater lies the volcano's fiery lava lake. A cooled blackened crust partially covers it like a dark canopy. Lava pushes through large holes in the canopy, bright orange and red spatters flying unpredictably and huge molten blobs bubbling up in slow motion. Close-up, observers describe the action as an addictive display of otherworldly fireworks. From afar, the volcano creates a ghostly red glow.

Since 1774 Marum has erupted at least 48 times. Strangely, the lava lakes never seem to fill and overflow the crater. Scientists suspect there may be a subterranean crack in the island that allows lava to leak into the sea. The volcano has proven deadly in the past, killing 10 people in 1894 and 21 in 1913. Today Marum continues to rumble, vents and fractures ooze steam, and pools of liquid fire are bubbling telltales of the island's molten heart.

The pit's malevolent red eye sits just a few hundred feet below the rim.

At the rim, Franck Tessier and Irène Margaritis peer into Marum. A great blast shaped the walls of the 1,200-foot-deep crater, tearing through two thick layers of lava.

Maori consider the three volcanoes *tapu,* meaning "sacred" and "sanctity."

T hree striking volcanoes make up Tongariro National Park in New Zealand. To the south looms the craggy mass of Ruapehu, at 9,176 feet the tallest peak on the North Island. It wakes every few years to expel enormous columns of steam and ash. To the north is Tongariro, the park's namesake, a sprawling complex of ancient craters where vents continuously exhale clouds of sulfurous gas. Between the two stands Ngauruhoe, forming a perfectly symmetrical cone, the epitome of an archetypal volcano. In 1887 the area became the nation's first national park and now encompasses nearly 200,000 acres. It is revered by the indigenous Maori people and is also a popular tourist destination. Oxidized iron and dark volcanic debris contrast with mineral-tinted Emerald Lakes, which visitors swim in despite its sulfurous odor. Perpetually steaming Red Crater is named for the chestnut-hued rock around its mouth. Below, footpaths wind across grassy hillsides punctuated by hissing vents. Tongariro's daunting summits and surrounding terrain served in part as the dark realm of Mordor for *Lord of the Rings* film director Peter Jackson, who invoked a bit of movie-making license. In reality it is a majestic region with stunning volcanic flare and an inspiring cultural history.

Mantled in winter white, 9,176-foot-tall Mount Ruapehu, foreground, reigns over Tongariro National Park in New Zealand, renowned for its volcanoes and glaciers. Sister peaks, the conical Ngauruhoe and broad Tongariro, rise in the distance. On Tongariro, below, oxidized iron and volcanic debris surround one of its Emerald Lakes.

HAWAII **Kilauea**

As 1600°F lava from Hawaii's Kilauea met the Pacific's waves in this July 2008 eruption, monstrous blasts flew hundreds of feet into the air. The ocean water boiled to vapor and spun off a 1,000-foot waterspout.

Kilauea has erupted continuously since 1983, adding nearly 600 acres to Hawaii's south shore.

"Kilauea molds the land, belching lava and fumes, hissing, roaring, always transforming. The view I photographed doesn't exist anymore."

Frans Lanting, National Geographic photographer

"It was like looking back to when the Earth was being born," says photographer Frans Lanting of this seething mix of smoke and gas rising from a spatter cone in Kilauea's Puʻu ʻŌʻō's crater.

Pacific Ocean
Kaua'i
O'ahu
Moloka'i
Lāna'i   Maui
Kaho'olawe
Hawaii   Hawai'i
(U.S.)   Kilauea
Pu'u 'O'o

stats
Hawaii: Kilauea

Type of volcano:
Shield volcano

Last significant eruption date:
2010

Elevation above sea level:
4,009 feet

Status:
Historical

Volcanic alert:
Hawaii's most active volcano during historical times has found its way into many Polynesian legends because of its explosive presence. A long-term eruption in 1983 produced lava flows that destroyed almost 200 homes.

Tourists gather at dusk, above right, to watch the newest land on Earth being formed by Kilauea's lava. When red-hot lava hits the ocean water, it often shatters into tiny fragments, creating black sand.

Local residents and visitors from around the world travel to Hawaii's Kilauea to pay tribute to Pele, the volcano goddess. They leave a wide variety of offerings including fake money, a raw pig's head, and baked chickens. In response, Kilauea continues to ooze, gush, and spew lava as it has done steadily since 1983 in an unmatched display of firepower.

The Big Island of Hawaii, which includes Kilauea and Mauna Loa volcanoes, has explosive—and relatively recent—origins. It was created less than a million years ago as magma blasted through Earth's crust, surging from an underlying hot spot as did the elder islands to its west and north. The hot spot continues to give birth: A seamount called Lo'ihi, its summit still thousands of feet below the sea surface, is now forming. Kilauea's current eruption has added more than 570 acres of land to the Big Island's southern coast.

Kilauea offers scientists an unprecedented natural laboratory. Representatives of the U.S. Geological Survey regularly measure temperatures and gases and analyze data from seismometers, tilt meters, and instruments linked to the global positioning system to measure earthquakes as well as land movement and deformation. Researchers venture dangerously close to the red-hot lava to collect samples.

Kilauea's landscape is filled with a bounty of volcanic features. Both of Hawaii's lava types occur on the volcano: the hot, fluid flows that harden into ropey blackened rock called *pahoehoe,* and the volcanic slag called *'a'a,* which solidifies into craggy rocks with brutally sharp edges. Porous lava nuggets glint green with the mineral olivine, and nests of volcanic glass called Pele's hair glimmer almost gold. Where sinuous rivers of lava break free, overlying gaseous clouds are tinged crimson. And mile-long lava tubes—formed when the outer surface of a flow crusts over, creating a conduit for the hot liquid inside—carry the most persistent of Kilauea's lava to its end, an explosive bath of cooling seawater.

The relatively gentle nature of lava flows in Hawaii creates great mounded mountains thought to resemble ancient round battle shields—giving them the name shield volcanoes. Kilauea rises slightly over 4,000 feet, dwarfed by its much larger neighbor, Mauna Loa, 13,681 feet high. Mauna Loa has been eerily quiet since 1984, and history suggests that an eruption is overdue.

This aerial view
of a 1983 eruption of
Kilauea's Pu'u Huluhulu
volcanic vent is part of
the volatile landscape
of Hawaii Volcanoes
National Park. Lava
from Kilauea builds
new land on Hawaii
island each year.

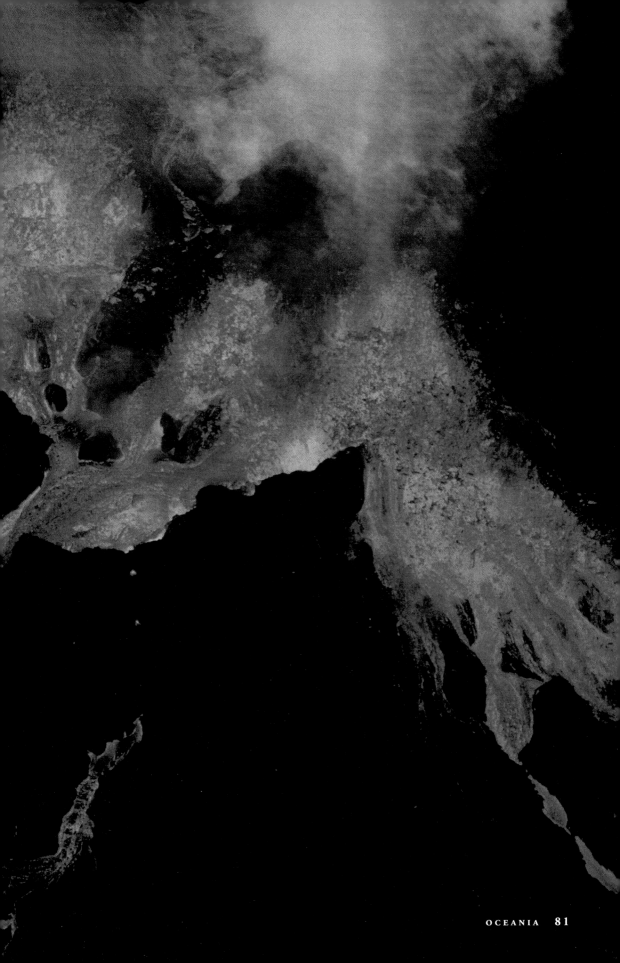

tallest active volcano
mt. damāvand
18,602 ft

deadliest eruption
mt. tambora
1815

tectonic location
eurasian, arabian, australian, and pacific plates

The stratovolcano Mount Mayon is not only the most active volcano in the Philippines, with some 50 eruptions in recorded history, it is also renowned for its perfect cone shape.

Merapi

Gas drillers probably
triggered an eruption
of mud from Merapi in
East Java in May 2006.
So far, the mud has
engulfed 12 villages
and displaced 10,000
families, and is still
surging today.

Inside the crater of
Mount Bromo on the
island of Java, men
hoist baskets to inter-
cept coins, vegetables,
and chickens thrown by
locals to appease the
resident spirit during
the festival of Kasada.

On Java alone, 120 million people live in the shadow of more than 30 volcanoes.

stats
Indonesia: Merapi

Type of volcano:
Stratovolcano

Last significant eruption date:
2007

Elevation above sea level:
9,737 feet

Status:
Historical

Volcanic alert:
One of Indonesia's most active volcanoes also resides just north of one of the world's most densely populated cities, Yogyakarta.

Deep in a trance, Baryo, a Hindu holy man, blesses offerings bound for Mount Bromo. Mired in poverty, many Indonesians petition the land's volcanic powers for a better life.

Indonesia is an archipelago of more than 17,000 islands straddling the western reaches of the Pacific Ring of Fire. Nowhere else on the planet do so many people live so close to so many active volcanoes—129 by one count. On Java alone, more than 120 million people live in the shadow of some 30 volcanoes, a proximity that has proved fatal to at least 140,000 people in the past 500 years. In 1815 the eruption of Tambora killed 92,000 people, and in 1883 the cataclysmic blast of Krakatau off Java's coast triggered a tsunami that claimed nearly 40,000 lives.

Java's Mount Merapi or "fire mountain" is one of the nation's natural-born killers. It rises almost 10,000 feet over forests and fields and is ranked among the world's most active volcanoes. Yet villages speckle the slopes of Merapi, where fertile land lures farmers into the danger zone. An eruption in 1930 killed more than 1,300; even in calmer times, plumes drift menacingly from the peak. Local hazards maps warn of pyroclastic and lava flows, rock falls, toxic gases, and glowing ejected rock fragments.

In May 2006, when the volcano's rumbling reached a crescendo, thousands fled Merapi's slopes and settled reluctantly into makeshift camps at lower altitudes a safer distance from the summit. Even the resident monkeys descended. It was well they did. The lava dome collapsed, and a scorching avalanche of rocks rushed down Merapi's western flank. Because of the evacuations, no lives were lost. Villagers who had not heeded the official warnings were fortunate to escape unharmed. They had looked for direction from Mbah Marijan, a shaman designated the gatekeeper of Merapi. It is his responsibility to perform rituals designed to appease an ogre spirit believed to inhabit the volcano's summit and to advise the citizens on all matters volcano. Bearing offerings, the faithful follow the gatekeeper to the top, where he placates the spirit within. In 2006, he advised his followers not to evacuate—and they got lucky.

Clearly modern science and mystical beliefs do not always see eye to eye, and one of the most difficult problems Indonesian authorities face is getting the local people to trust what the scientists say without appearing to disrespect traditional beliefs. Although Merapi is not one of the world's largest volcanoes, it threatens a staggering number of people living along its slopes. It and the region's other peaks are decidedly restless and will one day awake.

Scientists predicted
Mount Merapi's 2006
eruption, but some
mystically minded
locals refused to leave.
"The biggest problem
the Indonesians face,"
said a volcanologist,
"is getting the locals

Pinatubo

Roiling clouds of superheated ash surge from Mount Pinatubo in the Philippines, in June 1991, pumping two cubic miles of fine ash into the atmosphere during one of the 20th century's biggest volcanic explosions.

stats
Philippines: Pinatubo

Type of volcano:
Stratovolcano

**Last significant
eruption date:**
1993

**Elevation above
sea level:**
4,875 feet

Status:
Historical

Volcanic alert:
**The 1991 eruption
of Pinatubo was
one of the largest
in the 20th century.
It produced dramatic
pyroclastic flows and
left hundreds dead.**

**Pinatubo's thick ash
clouds turned three
days into nights in
nearby villages, above.
An acidic bright green
volcanic lake formed
in the crater, right,
both created by the
cataclysmic eruption.**

Mount Pinatubo had been a quiet, innocent-looking peak for some 600 years. Then in April 1991 steam eruptions, swarms of shallow earthquakes, increased sulfur dioxide emissions, and rapid growth of a lava dome heralded what was to be one of the century's most powerful and picturesque eruptions.

Up until then, many people in the region did not even realize that the peak was a volcano. Its heavily vegetated slopes and adjacent plains provided fertile soil and good living. Even the U.S. military chose the nearby flat ground to construct a large base. In all, nearly a million people were living in Pinatubo's shadow.

By early June 1991, an ominous rocky dome had swollen on Pinatubo's summit and began oozing lava. On June 12 the volcano blew in the first of what would be a series of devastating eruptions. Gas-charged magma breached the surface, exploded skyward, and fueled a mushroom cloud of gas and ash that eventually reached some 20 miles high and 200 miles wide; upper-level winds then began to spread it around the globe. In the nearby villages the sky turned dark; volcanic cinders and ash rained down. To compound the misery, Typhoon Yunya arrived with torrential rains that turned falling ash into a concrete-like slurry. Searing pyroclastic flows and huge avalanches of mud and debris surged off the volcano. At the summit the cataclysmic blast created a caldera more than a mile wide.

Despite the violence of the blast and the region's dense population, only 300 fatalities were recorded, most of which were caused by collapsing roofs. Much of the credit for the low death toll goes to scientists from the Philippine Institute of Volcanology and Seismology, who at the first hint of trouble stepped up efforts to monitor the volcano and assess the danger. They were later joined by colleagues from the U.S. Geological Survey. Working together with local authorities, they instigated the relocation of more than 25,000 people from the areas at greatest risk, and U.S. military personnel were evacuated from the base nearby.

Pinatubo's blast was so great that the global dispersal of gas and ash lasted for months and is now believed to have caused a worldwide average drop in temperature of 0.7°F that lasted two years. While the eruption illustrates the awesome power of volcanoes, it also shows how increased understanding can help avoid catastrophe.

The solitary, snow-covered peak of Mount Fuji, in Japan, rises above the fog. For centuries, Fuji has been Japan's holiest of natural sanctuaries.

stats
Japan: Fuji

Type of volcano:
Stratovolcano

Last significant eruption date:
1708

Elevation above sea level:
12,388 feet

Status:
Historical

Volcanic alert:
Although there hasn't been an eruption in several hundred years, low-frequency earthquakes were detected beneath the volcano in 2000 and 2001, causing some experts to think an eruption may be possible. The volcano lies 70 miles from Toyko.

A jellied bean-paste snack shaped like Mount Fuji, above right, is one of many such foods inspired by the iconic peak. Omnipresent in the culture, the volcano dominates the landscape, right.

More than 300 years have passed since Mount Fuji last erupted, and no one knows how long the lull will last. Sitting atop the Pacific Ring of Fire—a geologic seam of colliding tectonic plates that arc around the Pacific Ocean—Fuji has erupted at least ten times since the eighth century. Layer upon layer of lava and ash have helped make Fuji, at 12,388 feet, the tallest mountain in Japan.

On a clear day Fuji's flattened top, white in the winter and graphite gray in summer, can be seen from Tokyo some 70 miles away. Its solitary cone rises from Japan's boomerang-shaped main island of Honshu. Today the volcano is quiet, but during its last eruption in 1708, Fuji was anything but calm. A colossal earthquake, estimated today at magnitude 8.4, preceded the blast. Ash turned the sky dark and blanketed Tokyo, while on Fuji's southeast face a new crater was formed.

For the people of Japan, Fuji ("without equal") is more than a mountain or sleeping volcano, it is a source of national pride and a place to be revered. Fuji also has a place in popular culture, its image adorning a myriad of products. Each year some 400,000 people make the pilgrimage to Fuji to scramble to its summit, most coming during the summer climbing season. Some make the trek as a spiritual journey, while others relish the physical and mental challenge of the ascent.

In 2000 and 2001, scientists measured a swarm of small, shallow earthquakes that may have indicated the movement of magma, though their meaning remains unclear. It's not known when Fuji will next erupt, but today the consequences would surely be severe. Deadly gas and ash could spread over greater Tokyo, the biggest city in the world, where 35 million people now live and work. Local authorities are taking the dangers seriously and conduct evacuation drills in preparation. While rescue helicopters hover, loudspeakers blare, "Fuji has erupted! Forests are ablaze!" Many people have become complacent about the potential threat, but many others take it to heart and are planning for the worst, mapping out evacuation routes for their families.

These days the base of the sacred mountain is surrounded by the mundane—strip malls, fast-food shops, even a defunct Gulliver's Travels theme park complete with a giant Gulliver. But with its summit rising high above the clutter, the volcano stands majestically, quietly biding its time.

Fuji comes from early Chinese characters that mean "without equal."

An aerial view of Mount Fuji's crater is a gaping reminder of its last eruption in 1708. Over two miles above sea level, the crater is a half mile in diameter and 650 feet deep.

Daisetsuzan and Unzen

Daisetsuzan means "big snow mountain."

Fuji is not Japan's only mountain of concern. Also on the hot list are Unzen and the volcanoes of Daisetsuzan National Park. Uzen, on the island of Kyushu, has already taken a toll. In 1991 volcano experts Maurice and Katia Krafft went to film small pyroclastic flows bursting from the lava dome on Unzen. There they teamed up with Harry Glicken, a young geologist who had moved to Japan to be near frequent volcanic action. On the morning of June 3, accompanied by a group of Japanese journalists, they started up the volcano's slopes. They were never seen alive again. Later it was noted that a loud crack, like a thunderclap, had resounded off the mountain. A section of lava at the summit—gray on the surface but red hot underneath—had suddenly broken loose and crashed down the mountainside. The violent tumbling turned the lava into a fragmented cloud of killer heat that instantaneously killed the scientists and the journalists, along with waiting cab drivers and nearby farmers. In all, 43 lives were lost that day.

In Daisetsuzan National Park on Japan's northernmost island of Hokkaido, two massive white-capped volcanoes smolder, their smoking peaks rising above forested, river-washed slopes. The more southerly of the two, Tokachi-dake, last erupted in 2004. Steaming fumaroles punctuate the landscape, reminders of the underlying heat and danger of Japan's largest park.

Sulfurous fumaroles on Japan's Asahi Dake, left, mark the fiery heart of what was once a perfect cone. In 1991, Mount Unzen's violent eruption generated a massive pyroclastic flow, killing 43 and leaving a trail of destruction, below.

Kamchatka

Armed with guts and
a gas mask, French
explorer Franck Tessier
faces toxic steam from
one of 29 active vol-
canoes on Kamchatka
Peninsula—among the
most volcanically active
regions on Earth.

"Our water, our air, even our food tasted and smelled of sulfur."

Feodor Farberov, Russian guide to Kamchatka

Franck Tessier probes
fumaroles on the
toxic, sulfur-encrusted
firescape of an active
crater on Mutnovsky
volcano. Scorching
gases, uneven footing,
and falling slabs from
a collapsing glacier
lie ahead in the hellish
environment.

Bering Sea

Sea
of
Okhotsk

Kamchatka

Krasheninnikov
Uzon Caldera
Mutnovsky

Pacific
Ocean

stats
Russia:
Klyuchevskoy

Type of volcano:
Stratovolcano

**Last significant
eruption date:**
2010

**Elevation above
sea level:**
15,863 feet

Status:
Historical

Volcanic alert:
**Klyuchevskoy is the
highest and most
active volcano in the
Kamchatka Peninsula,
which is one of the most
active volcanic regions
on Earth. The peninsula
forms the northwestern
edge of the Pacific Ring
of Fire. Over a hundred
Kamchatka volcanoes
have erupted in the past
12,000 years.**

**Two overlapping stra-
tovolcanoes form twin
craters atop Krashenin-
nikov volcano, above
right. An Uzon caldera
mud pot is a slurry
of volcanic earth and
superheated water that
burbles in endlessly
changing patterns.**

R ussia's Kamchatka Peninsula lies along the Pacific coast a thousand miles north of Japan. It is one of the most remote and volcanically active regions on Earth. It also happens to be situated beneath the heavily traveled air routes between Asia and North America. Scientists and aviation experts carefully monitor the volcanoes of Kamchatka, which each year erupt and eject potentially dangerous clouds of ash high into the atmosphere.

Of the hundred-plus volcanoes dotting the 750-mile-long scimitar of land, 29 are active, and eruptions are frequent. Kronotsky volcano rises in a perfect cone to 11,575 feet; the most massive is Klyuchevskoy, which pours out an average of 60 million tons of basalt a year.

Bezymianny, one of a dozen volcanoes that make up the peninsula's Klyuchevskoy group, was thought to be dormant until 1955, when it began to rumble. On March 30, 1956, it exploded, producing a vast shroud of ash. Within two days the ash had reached Alaska and soon was detected over the British Isles. The blast flattened trees 15 miles away, and much like the Mount St. Helens 1980 blast, Bezymianny's big bang began with a giant avalanche. The volcano then blew out sideways, leaving a yawning horseshoe-shaped crater. Bezymianny has continued to erupt periodically ever since and today sits smoldering.

To the south is Mutnovsky. Along its slopes of wet slippery ash, fumaroles belch steam. The glaciers here are stained with cinders and ash. Underlying thermal activity causes the ice to break apart, creating crevasses and huge ice blocks that are constantly shifting. Craters walls are lined with red and yellow crystalline sulfur. In March 2000, blasts of steam rocked one of the craters, and the glacier inside it began to collapse. From the melted ice and snow was born a lake 650 feet across.

The Uzon caldera is a 40-square-mile depression on the Kamchatka Peninsula left behind by a cataclysmic eruption some 40,000 years ago. Here, boiling mud pools resemble bubbling pots of chocolate pudding, and shallow streams turn bright green from heat-loving algae. Milky blue lakes in the region are rich in carbon dioxide and highly acidic. Uzon's hot springs and fumaroles form the centerpiece of Kronotsky State Biosphere Reserve, which includes the Valley of Geysers, the second largest geyser field in the world after Yellowstone. The reserve is a stunning showpiece, encompassing 2.8 million acres of volcanic wilderness.

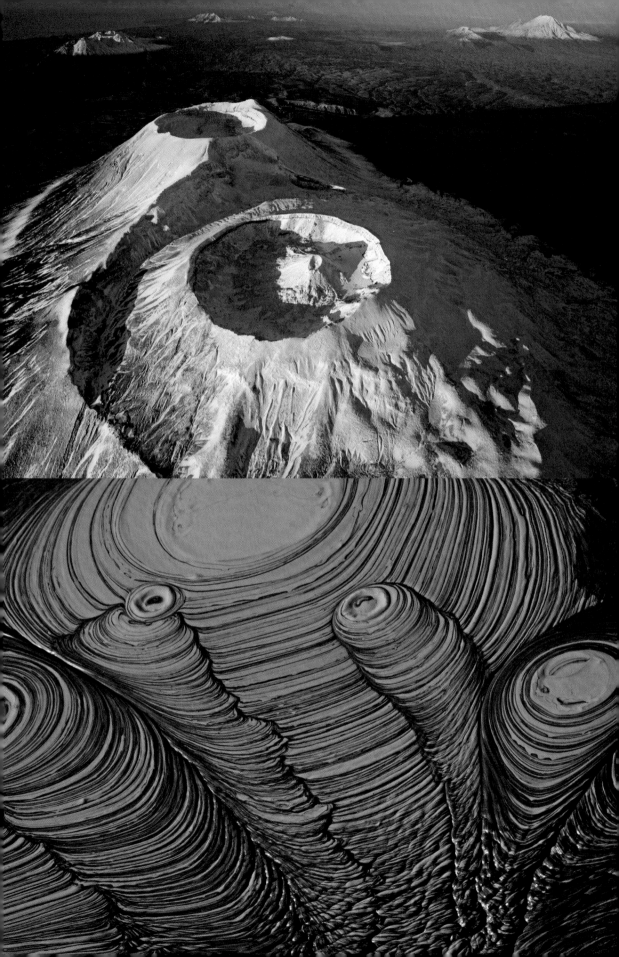

Gas bubbles rise from a thermal pond in Uzon caldera, where volcanism is dormant—for now. On the far shore a ranger armed with a rifle keeps watch for bears.

Kamchatka Peninsula is home to more than 100 volcanoes, 29 of which are active.

tallest active volcano
**mt. nos del salado
22,595 ft**

deadliest eruption
**mt. pelée
1902**

tectonic location
**north american,
cocos, nasca,
caribbean, south
american plates**

An infrared image of
Yellowstone's Pump
Geyser in action in the
western United States
reveals Earth's life-
blood at a boil. White
areas sizzle, reaching
over 200˚F. Blue means
wintry cool.

Redoubt

Like a nuclear cloud, hot ash rises in an updraft over a pyroclastic flow during a violent 1990 eruption of Alaska's Redoubt volcano; such glassy ash can stall aircraft engines when it melts onto the jet turbines.

Alaska
(U.S.)
Redoubt
Volcano
Arctic Circle
Mt. Katmai
Gulf of
Alaska

stats
Alaska: Redoubt

Type of volcano:
Stratovolcano

Last significant eruption date:
2009

Elevation above sea level:
10,197 feet

Status:
Historical

Volcanic alert:
In 2009, Redoubt sent an ash plume 50,000 feet above sea level, which eventually dusted communities 50 miles away. The eruption canceled flights in Anchorage, 100 miles to the northeast.

On March 21, 2009, Mount Redoubt awoke after an almost 20-year slumber. Here, steam rises from its summit crater. A day later, Redoubt erupted five times, sending an ash plume nine miles into the air.

I n December 1989, in a remote area of Alaska, Mount Redoubt began to show signs of unrest. It later erupted explosively, creating a towering cloud of gas and ash reaching at least 30,000 feet high. There were no casualties, and few people at the time took notice of the Alaskan pyrotechnics. But for the 245 people aboard a KLM jet flying through the region, it was a day they will never forget. When ash was sucked into the jet's engines, all four stalled. As the crew worked feverishly to restart the engines, the plane dropped some 14,000 feet, whereupon power was finally restored and the plane was able to land safely at Anchorage. It was an eye-opener for the industry and air-safety controllers and brought attention to the airborne dangers of volcanic ash clouds. On the ground, wildlife seemed little affected by the blast. Tracks in a new snowfall after the eruption revealed that a curious bear had lumbered to the crater's rim for a peek inside.

Mount Redoubt sits about a hundred miles southwest of Anchorage on the active northern rim of the Pacific Ring of Fire. The 10,000-foot peak began growing 800,000 years ago. Like a volcanic layer cake, it is constructed of alternating hardened lava flows and solidified ash and debris deposits. The volcano seems to go through episodes of extended activity followed by prolonged periods of quiet. For 22 years between the eruptions of 1966–68 and 1989–90, the mountain was still.

Redoubt remains an active volcano, erupting again in March 2009 with an explosive blast that this time sent ash and gas some 50,000 feet high. The volcano had started rumbling the previous year, and about a week prior to the big event, seismic activity, called volcanic tremor, told of magma on the move. Following the blast, a lava dome in the summit crater began to swell, reaching a volume estimated at 70 million cubic yards. Though this one remained stable, such domes, especially those that form on precariously steep slopes, can collapse, producing ash clouds and rock falls, hot debris flows, and rivers of mud.

The Alaska Volcano Observatory keeps a close eye on Mount Redoubt; in fact the mountain is probably one of the best watched, most monitored volcanoes in the world today. Judging from its recent history, scientists hope that it will provide ample warning before it blows the next time.

Nearly a century ago in 1912, in the Katmai area of southwestern Alaska, a long-quiet volcano came roaring to life, sending vast quantities of ash, pumice, and gas aloft. Multiple explosive blasts occurred over three long days; villagers a hundred miles away were thrown into darkness as the sky became black with ash. The National Geographic Society sponsored a series of expeditions to study the effects of the eruption. What they found was nothing short of astonishing. To the northwest was a valley of fumaroles, hot springs, and geysers unlike anything previously seen. The area soon became known as the Valley of Ten Thousand Smokes. In 1919 President Woodrow Wilson established Katmai National Monument, encompassing an area of 1,700 square miles, to preserve what was considered at the time to be among the greatest natural wonders of the world.

Discovered by botanist Robert Griggs in 1916, the Valley of Ten Thousand Smokes became a living laboratory for a 1918 Society expedition, left and right. The team trekked through suffocating vapors and across boggy earth to trap gases for analysis. Lacking wood, they cooked over natural steam. Today, most of the fumaroles have vanished.

Visitors in 1998 compare the mountain to its photo before its 1980 eruption. The volcano is one of some 20 in the Cascade Range connecting California and British Columbia.

Mount St. Helens came back to life from 2004 to 2008, growing a new lava dome in its crater.

Mount St. Helens, flanked by Mount Adams, background, is settling fitfully back into the volcanic landscape. Three decades ago its eruption killed 57 people and destroyed 230 square miles of forest.

stats
Washington: Mount St. Helens

Type of volcano:
Stratovolcano

Last significant eruption date:
2008

Elevation above sea level:
8,363 feet

Status:
Historical

Volcanic alert:
The volcano that erupted for nine hours in 1980 and created a mushroom-shaped column of ash that blew down or buried nearly 230 square miles of forest continues to be active today.

On May 18, 1980, a news photographer flying only three-quarters of a mile from the volcano's south side captured the paroxysmal awakening of Mount St. Helens about an hour after its 8:32 a.m. eruption.

N o one expected such a hellish, explosive blast. Scientists had warned that a menacing 320-foot bulge on Mount St. Helens summit was expanding five feet a day and might trigger a large avalanche or eruption. Yet no one could have predicted that the picturesque volcano would soon explode with the fury of a ten-megaton bomb. The 1980 eruption of Mount St. Helens was the largest in North America in more than six decades; it killed 57 people, flattened 230 square miles of forest, and filled riverbeds and valleys for miles with volcanic debris.

For over a hundred years, Mount St. Helens had been at peace, a scenic backdrop and recreational haven. In March of 1980, swarms of earthquakes announced the restless stirrings of magma and gas beneath the summit, and a few weeks later a bulge appeared on the mountain's northern flank.

Sunday May 18 began as a beautiful crisp day in the region, but at 8:32 a.m. a magnitude 5.1 earthquake struck. The mountain's bulge seemed to ripple, churn, and then it collapsed, triggering a volcanic cataclysm. A huge explosion ripped through the summit, and an enormous avalanche hurled volcanic debris downslope, moving north at speeds estimated as high as 150 miles an hour. Pent up gas and magma shot skyward in a searing spray of glassy ash and pulverized rock. Pressurized groundwater flashed into steam, and gas-filled rock exploded into dust, throwing 540 million tons of ash into the atmosphere. Blasts of hot ash, rock, and gas sent pyroclastic flows surging off the volcano, flattening trees like matchsticks and scalding the landscape for hundreds of miles. A wall of mud raged down the North Fork Toutle River and flowed into Spirit Lake. It was a disaster of unimagined proportions.

Thirty years later, throngs of tourists come to see Mount St. Helens's crater from a ridge named for geologist David Johnston, who was killed in the eruption. A 110,000-acre national monument has been established to let nature reclaim the land laid waste. Plants and animals have since returned to the region and grow more abundant each year. But there are no guarantees. In 2004 the volcano reawakened with several explosive but relatively minor blasts. For the next four years, a dome of lava swelled on the crater floor. Since then, Mount St. Helens has remained suspiciously quiet.

On May 18, 1980, Mount St. Helens self-destructed, setting off the biggest landslide in recorded history and losing 1,300 feet off its crown.

Buried in ash and washed out by mud slides, then controversially rebuilt in the 1980s and '90s for $160 million, State Route 504 leads to the Johnston Ridge Observatory, which overlooks Mount St. Helens's gaping crater.

Beneath Yellowstone lurks a beast unlike anything seen in human history. Steaming geysers, bubbling mud pools, and volcanic debris are evidence of its presence below. The oldest national park in the United States sits squarely atop one of the biggest volcanoes on Earth. Some 640,000 years ago it erupted in a blast that was a thousand times the size of Mount St. Helens's. A pillar of ash rose 100,000 feet, blanketing the West. Pyroclastic flows scorched and scoured the land. At 28 miles across, the collapsed caldera is difficult to discern, but the beast is very much alive. A swarm of earthquakes shook Yellowstone in 1985, again in 2008, and in early 2009. And since 2004, portions of the caldera have domed upward at nearly three inches a year, the fastest rate since detailed monitoring began in the 1970s.

Snowmelt and rainwater, superheated below the surface, blast from Yellowstone's Sawmill Geyser in a 20-foot-high geothermal belch. Hardened remains of old lava flows cover the area.

Sleeping Giant

Below Yellowstone, a hellish column of superheated rock—mostly solid, some viscous, some molten—rises from hundreds of miles within the Earth. Current stirrings may be remnants of a past eruption, or early harbingers of a still far-distant cataclysm.

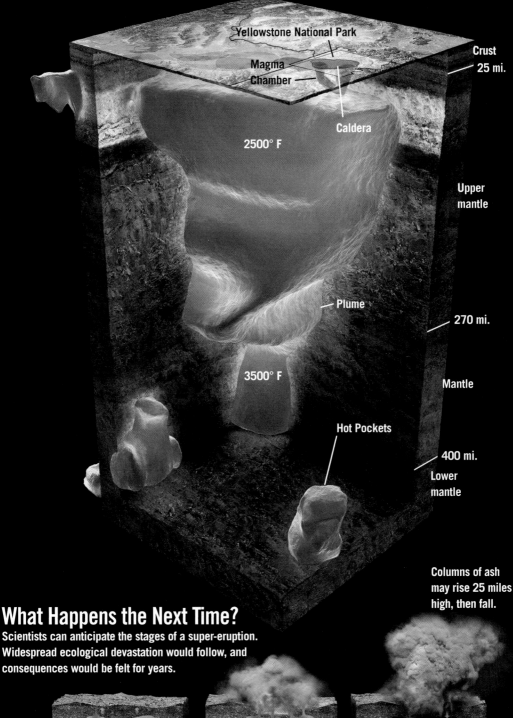

Yellowstone National Park

Magma
Chamber

Caldera

2500° F

Crust
25 mi.

Upper
mantle

Plume

270 mi.

3500° F

Mantle

Hot Pockets

400 mi.

Lower
mantle

Columns of ash
may rise 25 miles
high, then fall.

What Happens the Next Time?

Scientists can anticipate the stages of a super-eruption. Widespread ecological devastation would follow, and consequences would be felt for years.

Before the eruption

The Earth fractures

Eruptions continue

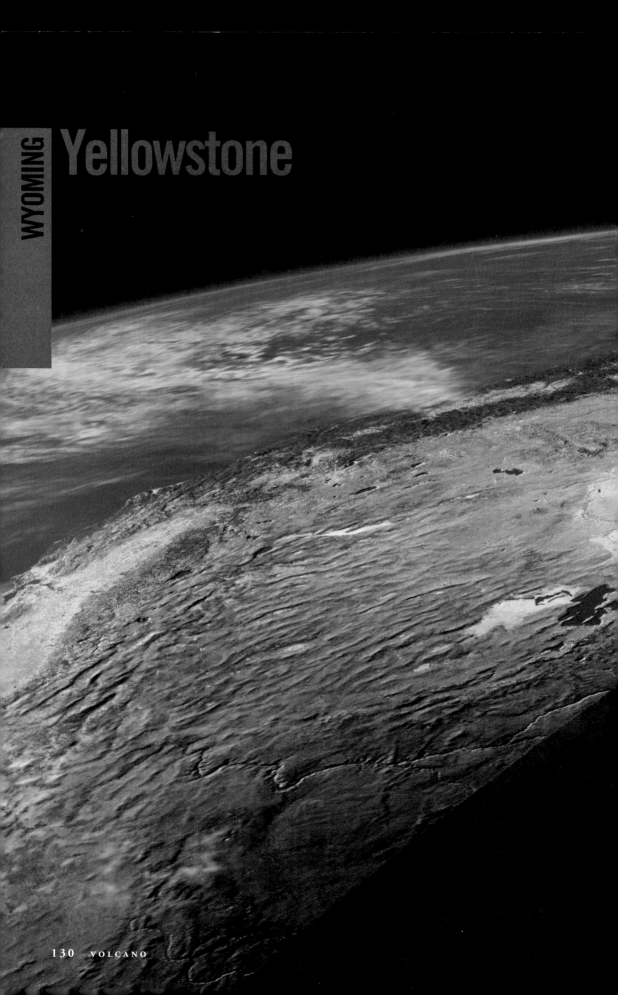

Yellowstone

Fire and debris spew from Yellowstone in an artist's depiction of a supervolcanic eruption, the end result of a monstrous plume of hot rock working up from deep inside the Earth. It's unlikely that Yellowstone's next major blast will occur in our lifetimes, but scientists have their ears to the ground.

Carefree children in
Santa Catalina Cuilote-
pec live in the shadow
of Mexico's sleeping
giant, Popocatépetl.
Stirring after more than
a half century of quiet,
it could soon erupt on
a scale not seen for a
thousand years.

Gulf of
Mexico
Tropic of Cancer

MEXICO
Parícutin
Popocatépetl

Pacific
Ocean

stats
Mexico: Popocatépetl

Type of volcano:
Stratovolcano

Last significant eruption date:
2010

Elevation above sea level:
17,802 feet

Status:
Historical

Volcanic alert:
Volcán Popocatépetl is Aztec for "smoking mountain"—suitable for the second largest volcano in North America, which towers over Mexico City some 40 miles away.

For centuries people have worshipped El Popo, above right, as a god who catches clouds and sends rain vital to their crops. Near the crater, below, the rainmaker, at left, leads a Catholic prayer at a gift-laden altar.

At the heart of Mexico stands a towering giant, one of the world's tallest and most dangerous volcanoes—Popocatépetl, at 17,802 feet. In 1993 the volcano began to stir after 70 years of calm, emitting gases, shaking the earth, and reaffirming its Aztec name "smoking mountain." Small explosions in the crater in 1994 prompted a chaotic evacuation of 25,000 people from the most vulnerable nearby villages. Since then El Popo, as Mexicans' affectionately call the peak, has intermittently thrown clouds of ash thousands of feet into the air and tossed glowing rocks down its steep upper slopes. But geologic evidence suggests it is capable of much more, and the consequences today would be devastating.

The farming communities and urban centers at the base of El Popo sit atop thick mudflow deposits. In some areas layers of pinkish pumice are buried beneath the ground. Both the mud and pumice are evidence of big eruptions in the past. Every thousand years or so, it appears that Popocatépetl lets loose a tremendous blast of hot ash and rock with fallout that blankets its own slopes as well as those of Iztaccihuatl, a dormant volcano ten miles to the north. In such eruptions, meltwater from two sizable glaciers on El Popo's north side fuels mudflows that race down ravines, burying everything in their path.

Today some 20 million people live within 50 miles of El Popo, and many of them—including the residents of Mexico City—have already been showered with ash. The most vulnerable, though, are the roughly 100,000 people residing in small villages at the very foot of the volcano.

According to old beliefs in the area, a volcano can be a god, a mountain, and a human all at the same time. People speak of El Popo with reverence, that he's angry when he erupts, or is unburdening himself when he ejects debris. Some believe he is the provider of food and water. Others deem an eruption to be a biblical penance that must be paid. For the less mystical, the risk of an eruption is simply worth the rich bounty the land provides. Faith and fatalism are both deeply rooted here.

Mexican scientists are monitoring El Popo, waiting and watching for signs of an impending eruption. Significant activity occurred in 1996, 2004, and 2005, and the volcano is expected to continue to throw out ash and rock occasionally. But with the passage of time, the probability of a catastrophic event grows ever stronger.

Popocatépetl rises above a church on a pre-Columbian pyramid in Cholula. One of the world's tallest active volcanoes at 17,802 feet, El Popo is also among the most dangerous.

Soufriere Hills

Soufriere Hills volcano, in Montserrat, puts on a pyrotechnic display in January 2010. With periodic violent activity since its devastating series of eruptions starting in 1995, it has become one of Earth's most closely monitored volcanoes.

"I was right here the first time it blow. The rock come falling down, but we weren't scared. We just stay here and watch the ash drift."

Gerard Dyer, Montserratian farmer

Smoke, steam, and ash billow from Soufriere Hills volcano as seen from Fort Ghaut, on Montserrat, on August 4, 1997. As the violent eruptions continued, the government ordered hundreds to evacuate a once "safe zone."

West Indies
Soufriere Hills
MONTSERRAT
(U.K.)
Caribbean Sea
SOUTH AMERICA

stats
Montserrat: Soufriere Hills

Type of volcano:
Stratovolcano

Last significant eruption date:
2010

Elevation above sea level:
3,002 feet

Status:
Historical

Volcanic alert:
Ash eruptions and pyroclastic flows from Soufriere Hills in 1995 forced evacuations and ultimately destroyed the capital city of Plymouth.

Monserrat's capital, Plymouth, was abandoned in 1996, above right. Pyroclastic flows from several major eruptions forced residents to evacuate before the city was covered in ash and mud, right.

I n July 1995, after almost four centuries of quiet, Soufriere Hills volcano awoke with an explosive start, disgorging a cloud of superheated steam, ash, and rock. It was the first in a series of unexpected eruptions that shocked the world and turned the lives of Montserratians upside down.

The tiny 39-square-mile island sits atop the Caribbean's version of the Ring of Fire, where the North American and South American plates are crashing into the Caribbean plate. Like its island neighbors, Montserrat was built by volcanoes, as evidenced by three old volcanic centers on Soufriere Hills. But after a long period of relative tranquility, people tend to forget about the inherent danger.

During the volcano's first series of eruptions in 1995, Plymouth, the capital and heart of the island, was blanketed in a thick cloud of ash. Within the crater the land began to bulge, as mudflows surged down the mountain's southeast flank. In the coming months, pyroclastic flows turned the tree-lined Tar River Valley into a bleak moonscape. Throughout the remainder of the year and into the next, the volcano generated repeated explosive eruptions. In July 1997 a major dome collapse released immense pyroclastic flows and surging ash-clouds; 19 people were killed. In the years since, the volcano has gone through cycles of relative quiet followed by dome growth, collapse, and eruption, spreading ash, mud, and searing flows across the island—most recently in February of this year.

On an island too small for one traffic light, there is not much room to run. About 5,000 people—nearly half the population—were evacuated when the eruptions first began. Hotels and restaurants closed, and cruise ships no longer stopped. Unemployment rose from 7 percent to 50 percent. Today, thousands more Montserratians have abandoned their homes, most emigrating to other Caribbean islands. Some have gone to Britain, which still holds Montserrat as a dependent territory.

Yet many people have stayed on, determined to maintain their way of life. It is, to say the least, a daunting task. The island is now subdivided into carefully monitored volcanic hazard zones, and access is dependent on the level of volcanic activity. No longer can people on Montserrat ignore their volcano, nor in fact does the rest of the world. Soufriere Hills has become one of the most closely watched volcanoes on the planet.

Hot clouds of gas, ash, and rock that raced down the side of Soufriere Hills volcano at nearly a hundred miles an hour turned the Tar River Valley into a wasteland.

Pacaya

On some days, fiery lava bombs shoot high above the summit.

Pacaya is among Guatemala's most active volcanoes, sitting just 17 miles from Guatemala City, the nation's capital and home to more than 2.5 million people. Since 1965 the 8,400-foot peak has been erupting nearly continuously, alternating between small explosive blasts and spattering, flowing lava. About a thousand years ago part of the volcano collapsed, sparking an avalanche of debris that spread for miles and created a large horseshoe-shaped caldera in which a new volcanic cone has grown. Just recently, in 1989, a major eruption produced a two-mile-high column of ash and widened the summit crater. Lava flows in the wake of the event continued for two years, damaging nearby villages.

The region's fiery peaks are the product of the subduction of the Cocos plate beneath the Caribbean plate, where melting of the underlying slab produces a plume of rising magma. Other areas of the world, such as Indonesia, have more active volcanoes than Central America, but in terms of sheer numbers of people living in harm's way, Central America with its large urban population is the unfortunate winner. Pacaya's eruptions today can often be seen from Guatemala City. For the most part, the activity is relatively benign and safe for viewing, luring both tourists and scientists alike.

Lava flows down the slope of Pacaya Volcano, left, while dormant Agua Volcano looms in the background. Gas, steam, and lava all play a part in an early morning eruption, below.

Arenal

On average, a volcano
erupts somewhere in
the world each week.

With a 5,000-foot-high conical peak surrounded by lush forests and an adjacent lake, Arenal is a postcard-worthy volcano that has paid off for Costa Rica, attracting thousands of tourists each year. At Arenal's base a ring of hot springs supplies baths for locals and elaborately piped pools for visitors. But the volcano, one of the country's most active, has not always been so benign.

In 1968, people in the area began to notice that their hot springs were getting hotter and that plumes of steam were emanating from the mountain. On July 29, three fissures opened on the volcano's west flank, sending blocks of steaming rock flying and scorching nearby villages with scalding ash and gas. Two days later another explosive blast occurred, and a cloud of ash and gas more than six miles high billowed over the volcano. Seventy-eight people died in the eruptions—from a volcano many people had thought extinct.

Since 1968 Arenal has been almost continually active, belching steam and fire, and putting on a great show. It's being carefully monitored, however, in case the show gets a little too hot for comfort.

Arenal Volcano puts on a fiery display at dusk, left. After lying dormant for hundreds of years, Arenal erupted in 1968 and has since earned the title of Costa Rica's most active volcano. At almost a mile across, Poas Volcano, below, has one of the largest craters on Earth.

An aerial photograph
captures a plume of
ash spewed by Chile's
Chaitén volcano in
May 2008. The cloud
spread across Chile to

"The most striking sensation of a volcanic eruption is its limitless force. That is the biggest force on earth . . ."

Haraldur Sigurdsson, volcanologist

Pacific
Ocean

SOUTH
AMERICA

CHILE

Chaitén

Atlantic
Ocean

stats
Chile: Chaitén

Type of volcano:
Caldera

Last significant eruption date:
2010

Elevation above sea level:
3,681 feet

Status:
Historical

Volcanic alert:
The 2008 eruption of Chaitén in southern Chile generated an ash cloud that disrupted air traffic and covered nearby towns in Argentina with debris. Bits of ash drifted as far as South Africa.

Chaitén blanketed the nearby town of Futaleufu with several inches of ash, above right, forcing residents to wear face masks—and even goggles, right—when venturing outdoors while most residents stayed indoors to avoid the suffocating ash.

U ntil 2008, only 42 of Chile's more than 100 volcanoes were thought to be active and highly dangerous. The count now stands at 43, as Chaitén has been added to the list. The peak seemed rather inconspicuous at the time, 3,700 feet high, topped with a 2.5-mile-wide forested caldera. But lurking inside the volcano's summit was a lava dome that had been growing for some 5,000 years. In May 2008 the dome collapsed and the volcano exploded.

The eruption shot ash and gas high into the stratosphere. Bolts of lightning flashed through the dark cloud of ash, putting on a fireworks display for the ages. Scalding pyroclastic flows surged off the mountain, accompanied by mudflows and falling cinders. Sitting just six miles away, the town of Chaitén was blanketed in ash and inundated with floods and mud. More than 4,000 people had to be evacuated by boat. Regional airports closed, and hundreds of flights were canceled in Chile and Argentina as the ash cloud spread, becoming visible even from space as it spread for hundreds of miles out over the Atlantic Ocean.

Though unexpected when it occurred, this was not Chaitén's first explosive eruption. Geologic evidence suggests that some 9,400 years ago the volcano cleared its pipes in a massive blast that also produced fiery pyroclastic flows and ashfalls. Chaitén is one in a series of volcanoes that line the southwestern coast of South America, a result of the collision between the Nasca and South American plates.

Following the 2008 eruption, a new lava dome began to grow. Its swelling has been accompanied by numerous eruptions, which are ongoing today. Plumes of gas and ash periodically burst into the sky, and mudflows, pyroclastic surges, and earthquakes continue to plague the area. Chileans are not the only ones worried about the feisty volcano. Neighboring Argentina and the airline industry are keeping a close eye on it as well. Chaitén's ash clouds can reach well above a jet's normal cruising altitude; in the first week of the 2008 eruption, several airplanes sustained damage. Along with disruption to air and ground travel, ash from Chaitén has the potential to impact the health of people living in both Chile and Argentina.

The one piece of good news is that the eruption brought new attention to the dangers of Chilean volcanoes and inspired a national monitoring program that will keep tabs on the country's highest risk peaks.

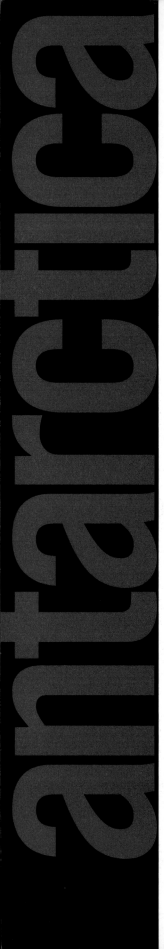

antarctica

tallest active volcano
mt. erebus
12,447 ft

latest eruption
2010

tectonic location
antarctic
plate

Steam drapes the cliffs on Bellingshausen as waves eat away at rock born from the crater's fire. Stoking that inferno is one of Earth's fastest tectonic plates, which moves below at nearly three inches a year.

Erebus

Created by steaming fumeroles that freeze on contact with the Antarctic air, a soaring ice tower is one of hundreds that form on the flanks of Mount Erebus. Part of the Pacific Ring of Fire, Erebus is the southernmost active volcano on Earth.

Sources

Material used in this book is drawn from the following
National Geographic Society books and magazine articles:

National Geographic books:
Raging Forces: Life on a Violent Planet, 2007

National Geographic magazine articles:
Volcanoes of Alaska, August 1912
The Ten Thousand Smokes Now a National Monument, April 1919
*In Iceland's Fourteenth Volcanic Eruption of This Century, a Mountain is Born, and
  a Village Fights for Its Life*, July 1973
In the Path of Destruction, January 1981
Africa's Great Rift, May 1990
The Blast Not Heard Around the World, December 1990
Volcanoes: Crucibles of Creation, December 1992
Montserrat: Under the Volcano, July 1997
Popocatépetl: Mexico's Smoking Mountain, January 1999
Mount St. Helens: Nature on Fast Forward, May 2000
Inside the Volcano, November 2000
Etna Ignites, February 2002
Russia's Frozen Inferno, August 2001
Fuji: Japan's Sacred Summit (Except When It's Not), August 2002
Ol Doinyo Lengai, January 2003
Red Hot Hawaii: Volcanoes National Park, October 2004
Vesuvius: Asleep For Now, September 2007
Gods Must Be Restless: Living in the Shadow of Indonesia's Volcanoes, January 2008
Daisetsuzan, August 2008
Let It Be: Russia's Kronotsky Reserve, January 2009
Between Fire and Ice: New Zealand's Tongariro National Park, July 2009
When Yellowstone Explodes, August 2009
Mountain Transformed: Mount St. Helens, May 2010

Acknowledgments

Jim Kauahikaua, USGS
Judy Fierstein, USGS
Cynthia Gardner, USGS
Dina Venezky, USGS
Chris Waythomas, USGS
Tina Neal, USGS
John Pallister, USGS
Clarice Ransom, USGS
Thomas L. Murray, USGS
John C. Eichelberger, USGS
David Applegate, USGS
Tim Dixon, University of Miami
Grace Hill, Associate Managing Editor
Samantha Foster and Matt Propert, National Geographic contributors

Illustrations Credits

Cover, Arctic-Images/Corbis; 2-3, **Carsten Peter**/NG Stock; 4-5, **Chris Johns**/NG Image Collection; 7, **Warren Goldswain**/Shutterstock; 8-9, **John Stanmeyer**/NG Stock; 10-11, **Carsten Peter**/NG Stock; 12-13, **Carsten Peter**/NG Stock; 14, NG Books; 15, **Holger Mette**/Shutterstock; 16, **George Burba**/Shutterstock; 16-17, **Jaime Quintero**/NG Image Collection; 17, **Carol Schwartz;** 18-19, World Map Sources: Holocene volcano data–Global Volcanism Program/Smithsonian Institution, Bathymetry–CleanTOPO2; 21, **Peter Vancoillie**/YOUR SHOT; 22-23, Reuters/**Lucas Jackson;** 24-25, **Omar Oskarsson**/AFP/Getty Images; 26, NG Books; 27 (upper), **Peter Muhly**/AFP/Getty Images; 27 (lower), **Joerg Koch**/AFP/Getty Images; 28-29, Reuters/**Ingolfur Juliusson;** 30, **Emory Kristof**/NG Image Collection; 31, **Robert S. Patton**/NG Image Collection; 32-33, **Robert Clark**/NG Image Collection; 34, NG Books; 35, **James L. Stanfield**/NG Stock; 36-37, NGM Maps; 37, **Mikhail Nekrasov**/Shutterstock; 38-39, U.S. Army Air Forces; 40-41, **Carsten Peter**/NG Stock; 42-43, **Carsten Peter**/NG Stock; 44, NG Books; 45, **Carsten Peter**/NG Stock; 46-47, **Carsten Peter**/NG Stock; 49, **Vulkanette**/Shutterstock; 50-51, **Carsten Peter**/NG Stock; 52-53, **Carsten Peter**/NG Stock; 54, NG Books; 55, **Carsten Peter**/NG Image Collection; 56-57: **Carsten Peter**/NG Image Collection; 58, **George Steinmetz**/NG Stock; 59 (both), **Carsten Peter**/NG Stock; 60-61, **Chris Johns**/NG Image Collection; 63, **Pichugin Dmitry**/Shutterstock; 64-65, **Carsten Peter**/NG Image Collection; 66-67, **Carsten Peter**/NG Image Collection; 68, NG Books; 69 (both), **Carsten Peter**/NG Image Collection; 70-71, **Carsten Peter**/NG Image Collection; 72, **Stuart Franklin**/NG Stock; 73, **Stuart Franklin**/NG Stock; 74-75, **Steve** and **Donna O'Meara**/NG Stock; 76-77, **Frans Lanting**/NG Stock; 78, NG Books; 79 (upper), **Steve** and **Donna O'Meara**/NG Stock; 79 (lower), **Sarah Leen**/NG Stock; 80-81, **James L. Amos**/NG Image Collection; 83, **Ewan Chesser**/Shutterstock; 84-85, **John Stanmeyer**/NG Stock; 86-87, **John Stanmeyer**/NG Stock; 88, NG Books; 89, **John Stanmeyer**/NG Stock; 90-91, **John Stanmeyer**/NG Stock; 92-93, **Alberto Garcia**/Corbis; 94, NG Books; 95 (upper), **Philippe Bourseiller**/Photo Researchers, Inc.; 95 (lower), **Steve** and **Donna O'Meara**/Photo Researchers, Inc.; 96-97, **Karen Kasmauski**/NG Stock; 98, NG Books; 99 (both), **Karen Kasmauski**/NG Stock; 100-101, **Karen Kasmauski**/NG Stock; 102, **Michael S. Yamashita**/NG Stock; 103, **James L. Stanfield**/NG Image Collection; 104-105, **Carsten Peter**/NG Stock; 106-107, **Carsten Peter**/NG Stock; 108, NG Books; 109 (both), **Michael Melford**/NG Stock; 110-111, **Carsten Peter**/NG Stock; 113, **Mark Thiessen**/NGS Image Collection; 114-115, AP Photo/Alaska Volcano Observatory/USGS/**R. Clucas;** 116, NG Books; 117, AP Photo/Alaska Volcano Observatory/USGS/**R. Clucas;** 118-119, **Robert F. Griggs**/NG Image Collection; 118, **Robert F. Griggs**/NG Image Collection; 119, **Frank L. Jones**/NG Image Collection; 120-121, **Jim Richardson**/NG Stock; 122-123, **Diane Cook** and **Len Jenshel**/NG Stock; 124, NG Books; 125, **Roger Werth**/The Daily News (Washington); 126-127, **Diane Cook** and **Len Jenshel**/NG Stock; 128, **Mark Thiessen**/NG Image Collection; 129, **Hernán Cañellas**/NG Image Collection;130-131, **Hernán Cañellas**/NG Image Collection; 132-133, **Sarah Leen**/NG Stock; 134, NG Books; 135 (both), **Sarah Leen**/NG Stock; 136-137, **Sarah Leen**/NG Stock; 138-139, AP Photo/**Wayne Fenton;** 140-141, AP Photo/**Kevin West;** 142, NG Books; 143 (both), **Vincent J. Musi**/NG Stock; 144-145, **Vincent J. Musi**/NG Stock; 146, **Olivier Voisin**/Photo Researchers, Inc.; 147, **Bobby Haas**/NG Image Collection; 148, **Mark Vincent Mueller**/YOUR SHOT; 149, **Gerry Ellis**/Minden Pictures/NG Stock; 150-151, **Alvaro Vidal**/AFP/Getty Images; 152, NG Books; 153 (upper), **Christian Brown**/AFP/Getty Images; 153 (lower), AP Photo/**Ferran Majol**; 155, **Maria Stenzel**/NG Stock; 156-157, **Maria Stenzel**/NG Stock.

Volcano

The National Geographic Society is one of the world's largest nonprofit scientific and educational organizations. Founded in 1888 to "increase and diffuse geographic knowledge," the Society works to inspire people to care about the planet. It reaches more than 325 million people worldwide each month through its official journal, *National Geographic,* and other magazines; National Geographic Channel; television documentaries; music; radio; films; books; DVDs; maps; exhibitions; school publishing programs; interactive media; and merchandise. National Geographic has funded more than 9,000 scientific research, conservation and exploration projects and supports an education program combating geographic illiteracy. For more information, visit nationalgeographic.com.

For more information, please call
1-800-NGS LINE (647-5463)
or write to the following address:

National Geographic Society
1145 17th Street N.W.
Washington, D.C. 20036-4688 U.S.A.

Visit us online at www.nationalgeographic.com

For information about special discounts for bulk purchases, please contact National Geographic Books Special Sales: ngspecsales@ngs.org

For rights or permissions inquiries, please contact National Geographic Books Subsidiary Rights: ngbookrights@ngs.org

ISBN: 978-1-4262-0761-7

Printed in USA

10/CML-CK/1

Nina D. Hoffman
Executive Vice President;
President, Book Publishing Group

Barbara Brownell Grogan
Vice President and
Editor in Chief

Marianne R. Koszorus
Director of Design

Carl Mehler
Director of Maps

R. Gary Colbert
Production Director

Jennifer A. Thornton
Managing Editor

Meredith C. Wilcox
Administrative Director,
Illustrations

Susan Blair
Project Editor/
Illustrations Editor

Bridget A. English
Editorial Assistant

Bob Gray
Designer

Robert L. Booth
Text Editor

Julie Beer
Researcher

Al Morrow
Design Assistant

**XNR Productions
and Michael McNey**
Map Research and Production

Lisa A. Walker
Production
Project Manager

Rachel Faulise
Manager, Manufacturing
and Quality Management